Dinosaurs
and their living relatives

Second edition

Contents

Preface to first edition

Could dinosaurs – animals that died out 65 million years ago – be closely related to crocodiles and birds alive today? After reading this book, you should be able to decide for yourself.

The book takes a completely new approach to the study of dinosaurs. It sets out to discover how dinosaurs are related to other animals – both living and extinct. It begins by explaining a simple method for working out the relationships between animals. Then, using many photographs and diagrams, it applies this method to the dinosaurs. The book ends with a unique series of new full-colour illustrations of many of the Natural History Museum's most famous dinosaurs – as they may have appeared when they were alive.

Dinosaurs and their living relatives has been produced in conjunction with a permanent new exhibition of the same name that opened at the Natural History Museum in May 1979. This exhibition, which was planned with the guidance of Museum experts, is the third in the Museum's major new exhibition programme.

Preparing the exhibition and its companion book has involved the effort and imagination of a great many people, both within the Museum and outside, and I should like to take this opportunity of thanking everyone concerned.

R.H. Hedley, Director
British Museum (Natural History)
August 1979

Dinosaurs, from left to right:
Iguanodon, Altispinax, Polacanthus, Hypsilophodon (two).

3

Dinosaurs, from left to right:
Camptosaurus (two), *Stegosaurus,*
Brachiosaurus, Apatosaurus.

4

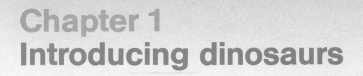

Chapter 1
Introducing dinosaurs

Dinosaurs dominated the land for 140 million years. They became extinct 65 million years ago—63 million years before the first human beings appeared.

How are dinosaurs related to animals alive today?

In this book, you can find out how to work out relationships between animals. And, as you go through the book, you will use this information to discover which of today's animals are closely related to dinosaurs.

Who were the dinosaurs?

The word 'dinosaur' means terrible lizard. Richard Owen, who was to become the first Director of the British Museum (Natural History), gave the name to some huge fossil creatures that were discovered last century – creatures such as *Megalosaurus* and *Iguanodon*. Since then, many more dinosaurs have been discovered – and, as you can see, they are very varied. Indeed, a scientist today would probably not group them all together.

What were dinosaurs like?

Some ate meat . . .

Tyrannosaurus was a fierce, flesh-eating dinosaur with large, sharp teeth and powerful jaws.

about 12 metres long

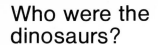

Some were slow and clumsy . . .

Stegosaurus weighed 2 tonnes, and had a brain about the size of a walnut.

about 6 metres long

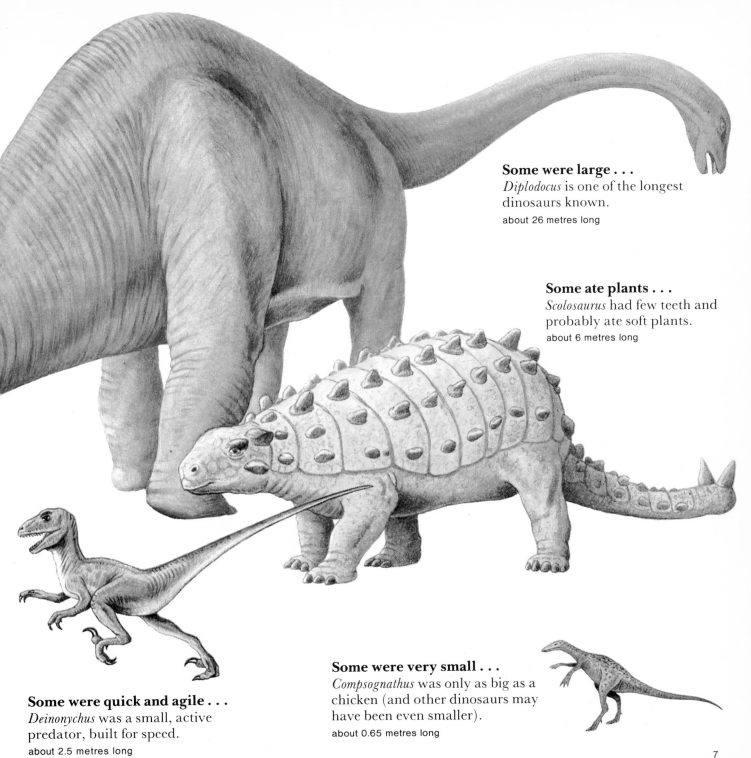

Some were large . . .
Diplodocus is one of the longest
dinosaurs known.
about 26 metres long

Some ate plants . . .
Scolosaurus had few teeth and
probably ate soft plants.
about 6 metres long

Some were quick and agile . . .
Deinonychus was a small, active
predator, built for speed.
about 2.5 metres long

Some were very small . . .
Compsognathus was only as big as a
chicken (and other dinosaurs may
have been even smaller).
about 0.65 metres long

7

Learning about dinosaurs

There are no dinosaurs alive today. All we have are their fossilized remains – and many of these are incomplete. So how can we 're-create' dinosaurs as you saw them in the last few pages? How do we know what they looked like, how they moved or what they ate?

The answer is that we don't know for certain. We can only make an 'informed guess' about what dinosaurs were like, and we do this with the help of knowledge that we have of animals alive today. For example, we can compare structures such as horns, from a present-day rhinoceros and *Triceratops*, and suggest that *Triceratops* used its horns to defend itself.

In the same way, we can compare the teeth of *Tyrannosaurus* with those of a living animal, such as a tiger, and guess that it too was a meat eater.

The Maidstone *Iguanodon*

This is one of the first dinosaur fossils ever found – the bones of an *Iguanodon*. It was discovered in a quarry in Kent in 1834, and bought for Dr Gideon Mantell, who named the dinosaur.

Supposing you had discovered the *Iguanodon* bones yourself? How would you fit them together?

(Remember, when the fossils were found, no-one had ever seen a complete dinosaur skeleton, or even knew that dinosaurs existed.)

You could juggle the bones around until you got a rough fit . . .

But then you might end up with something like this!
(In fact this was how some fossil elephant bones were fitted together, at a time when people believed in unicorns.)

You could use your knowledge of present-day animals to help you sort out the bones . . .

You might start by identifying the backbone, and the main limb bones, and put these together. (On the fossil, the different bones have been numbered. For example, numbers 1 and 2 are hind limb bones.)

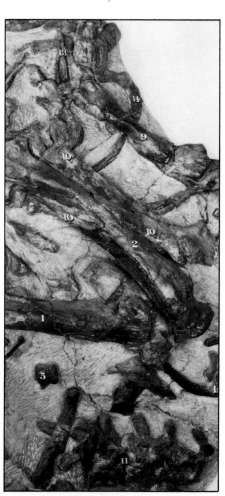

You could then make an 'informed guess' about what the animal looked like . . .

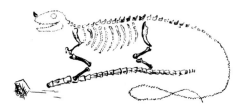

This is Mantell's original reconstruction of *Iguanodon*. He gave it its name ('iguana tooth') because its teeth were like those of a modern iguana. And he *guessed* that the extinct creature looked like a giant lizard. (One of the things he did wrong was putting the thumb bone on the nose, like a little horn!)

ISBN
0800105 50

stair

s?

AACP

This is an 1854 reconstruction of *Iguanodon*. The thumb bone was still wrongly placed on the nose, and the skeleton was assembled on all fours.

Twenty three years later, a huge collection of complete *Iguanodon* skeletons was discovered – and scientists realized that the animal usually walked on two legs, not four as they had originally thought.

But skeletons alone do not provide enough information for scientists to produce a detailed reconstruction like this one – which was made at the Natural History Museum in 1979. Scientists also have to rely on their knowledge of the anatomy of present-day animals to make 'informed guesses' about things like the muscles, skin and colouring of fossil animals.

In the rest of this book, you will see how we use our knowledge of fossil and present-day animals to work out the relationships between them.

Chapter 2
Working out relationships

How can we work out
relationships between living things?

All living things are related to one another. They are all descendants of a single ancestor that evolved on Earth millions of years ago.

We cannot find out how living things are related to each other by going back in time, so we must find another way of learning about relationships.

We can start by grouping similar living things together.

Grouping is something that we do automatically in our daily lives. We recognize that some things have more in common than others, and we put them into groups accordingly.

Try making some groups for youself.

Look at the seven animals on the right. Now try sorting them into the following groups:

1 Animals with backbones

2 Animals with four bony limbs

3 Animals with feathers

Turn the page to check your answers.

eagle

snail

lizard

tiger

cod

duck

toad

Were your groups the same as these?

Look again at the groups.

Did you notice that each new group is made up of only *some* of the members of the previous group? At each stage, you grouped together fewer animals—but they had more in common.

The three groups can also be shown like this—one inside the other.

animals with feathers

animals with four bony limbs

animals with backbones

How does grouping help us?

● It provides us with a kind of filing system—a way of organizing our knowledge.

Look back at the pictures on page 12. Can you fit any of the other animals into the groups above?

● It can also help us to work out what other features an organism may have. We know, for example, that a peacock has feathers—so we also assume that, like all other birds, it has four bony limbs and a backbone. Informed guesses like these are very important in interpreting fossils, such as dinosaur skeletons. Often, only a few features have been preserved and a great deal of guesswork is involved.

You grouped the animals together according to features that they share. The features you used are ones that show the same basic structure or pattern. Such features are called **homologues**. We think that the reason such features are similar is because they have been inherited from the same ancestor—the **common ancestor** of the group.

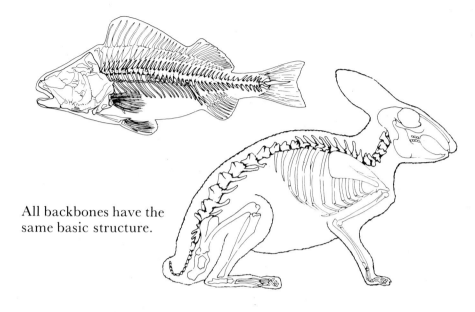

All backbones have the same basic structure.

king penguin
Aptenodytes patagonica

greater bird of paradise
Paradisaea apoda

No matter how different they look, all feathers have the same basic structure.

Not all homologues are easy to spot—sometimes we have to examine the features closely to discover their underlying similarities.

For example . . .

All animals with four bony limbs have the same basic pattern of bones in their limbs. This kind of limb is known as a **pentadactyl limb**. ('Pentadactyl' means five-fingered.)

You can see that, in many of the limbs, not all the five 'fingers' are present—the limbs have been greatly modified and perform different functions.

(Because the limbs look very different, the bones have been coloured to show the pattern more clearly.)

Madagascar flying fox
Pteropus medius

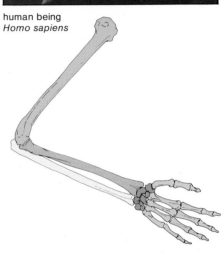

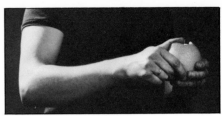

human being
Homo sapiens

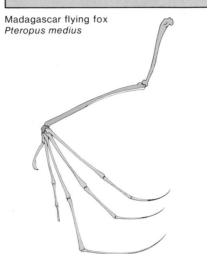

herring gull
Larus argentatus

wild sheep (mouflon)
Ovis musimon

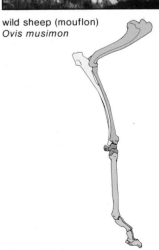

Misleading similarities

Some features look very similar in different animals because these features have a similar job to do. But, when we look closely, we can see that the features are really quite different. They do not share the same basic structure or pattern, and the similarity is only superficial.

Such features are called **analogues**. Analogues can be very misleading. They cannot tell us if different animals share the same ancestor, and biologists do not use them to form groups.

spur-thighed tortoise
Testudo graeca

hairy armadillo
Chaetophractus nationi

Tortoises and armadillos both have shells to protect them, but their shells are constructed quite differently. The ribs form part of the shell in tortoises, but in armadillos they do not.

We have legs, and so do locusts, but the two structures are really quite different. Our muscles are attached to the outside of our skeletons, whereas locusts have their muscles *inside* their skeletons.

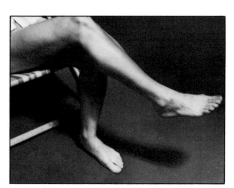

human being
Homo sapiens

desert locust
Schistocerca gregaria

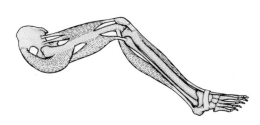

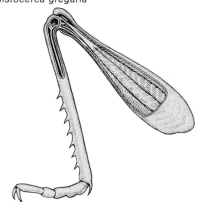

Can you recognize homologues?

Two of these animals have structures that are homologous to a bird's wing. Which do you think they are?

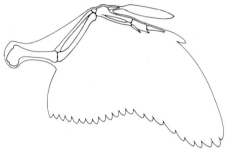

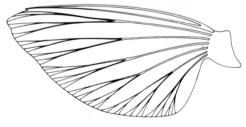

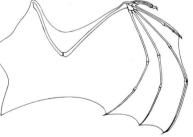

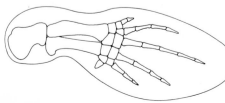

The bat and whale have front limbs that are homologous to a bird's wing—the arrangement of the bones is the same.

The fly's wing has no bones at all, and the bony rays in the flying fish wing are not homologous to the bones in a bird's wing.

18

We assume that animals that share a common feature are related to each other, through a common ancestor.

The more **homologues** a group of animals share, the more closely we assume they are related.*

Thus, all animals with backbones are related.

Within the group of 'animals with backbones', all those with four bony limbs are more closely related to one another than they are to other members of the group.

And all those four-bony-limbed animals that have feathers are more closely related still.

* In this book we define closeness of relationship in terms of **recency of common ancestry**. Feathers evolved more recently than four bony limbs, so we say that all animals with feathers are more closely related to one another than to any other animal with four bony limbs.

Clades

A group of animals that share a
unique homologue is called a
clade.

Many familiar groups are clades . . .

What makes a bird a bird?

It has feathers.

No other animal has feathers.
All birds share this unique
homologue.

Birds form a clade.

What makes a mammal a
mammal?

It has mammary glands, and
suckles its young. No other animal
has mammary glands.
All mammals share this unique
homologue.

Mammals form a clade.

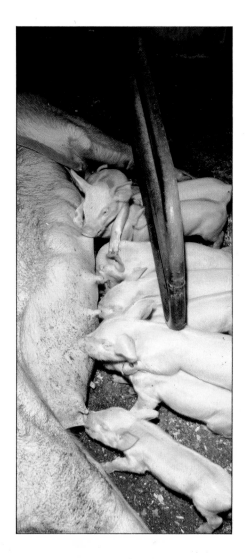

But some familiar groups are not clades . . .

We group fishes together because they share a number of homologues. But each of the homologues is shared by other animals as well.

What makes a fish a fish?

It has scales . . .

. . . but crocodiles do too.

It has a backbone . . .

. . . Do any other animals have a backbone?

It has gills . . .

. . . but so do tadpoles.

It has jaws . . .

. . . Do any other animals have jaws?

We do not know of any homologue that *only* fishes have.

So fishes do not form a clade.

The method grouping animals into clades – called **cladistics** – is only one of the many methods used in classification today. We have chosen to use cladistics in this book because clades can be clearly and simply defined.

You have seen that some traditional groups, such as the fishes, are not clades. As a result there is a conflict between the traditional system of classification, which many scientists still find useful, and cladistics.

Patterns of clades

To show how different clades are related to one another, we usually draw a branching diagram, called a **cladogram**.

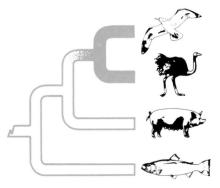

The seagull and the ostrich are linked because they both have feathers . . .

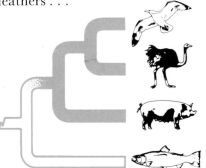

. . . and they are linked with the *pig* because all three have four bony limbs . . .

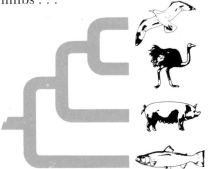

. . . and these are all linked with the *salmon*, because they all have a backbone.

The cladogram is equivalent to the grouping diagram you met on page 14.

You may have met similar diagrams, called 'set' diagrams, in mathematics.

Where do the dinosaurs fit in?

Here are the four clades you have met so far. Which ones do the dinosaurs belong to?

If you want to try to decide for yourself, cover up the right-hand column, which gives the answers.

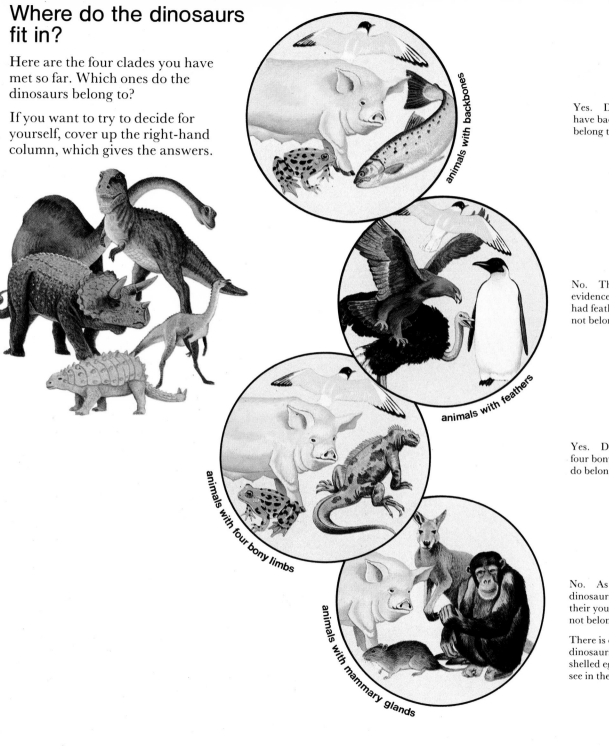

animals with backbones

animals with feathers

animals with four bony limbs

animals with mammary glands

Yes. Dinosaurs do have backbones, so they belong to this clade.

No. There is no evidence that dinosaurs had feathers, so they do not belong to this clade.

Yes. Dinosaurs have four bony limbs, so they do belong to this clade.

No. As far as we know, dinosaurs did not suckle their young. So they do not belong to this clade.

There is evidence that dinosaurs laid hard-shelled eggs, as you will see in the next chapter.

Chapter 3
Dinosaurs and their relatives

You have seen that dinosaurs belong to the clade of animals with four bony limbs. But *which* members of this clade are they most closely related to?

To find out, we must look for a homologue that dinosaurs share with only some members of the group.

The present is the key to the past.

We know more about living animals than about dinosaurs. So we start by taking a look at how living animals are related to each other . . . then we try to fit in the dinosaurs.

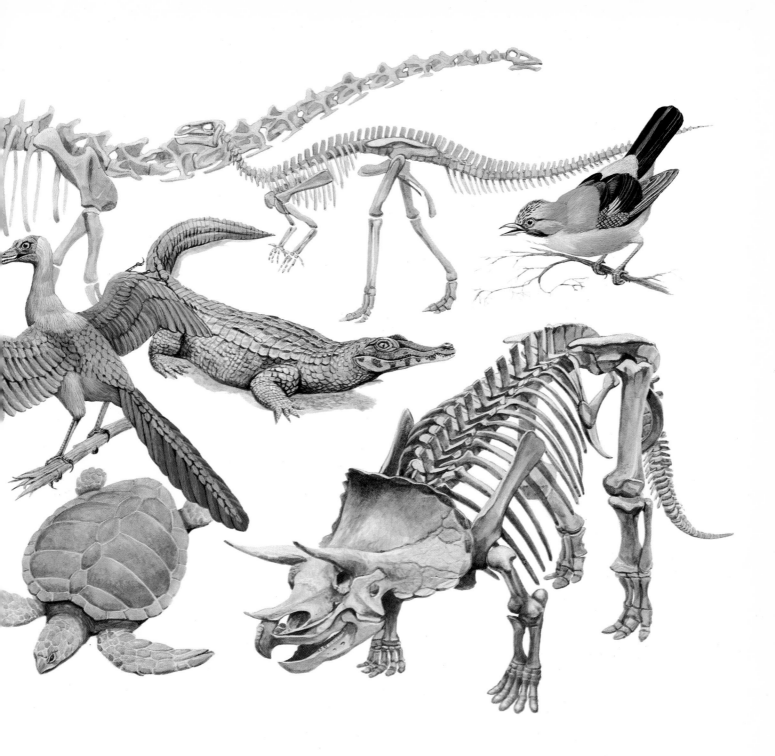

These animals may look very different . . .

. . . but they are all closely related.

When we look at the way they develop before they are born, we discover that these animals have an important feature in common. In all of them, the embryo is surrounded by a membrane called an **amnion**. This membrane forms a sac, filled with fluid which cushions the embryo and keeps it wet.

Because they all share this important homologue, we group all these kinds of animals together in a clade. We call them **amniotes**. (One other living animal – the tuatara – also has an amnion. We think that the tuatara is closely related to lizards and snakes.)

birds

bird embryo*

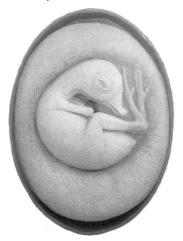

crocodiles

crocodile embryo*

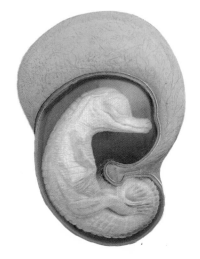

lizards and snakes

lizard embryo*

turtles and tortoises

turtle embryo*

mammals

mammal embryo

* These embryos are contained in a shell.

All embryos must be protected in some way, but if the eggs are laid in water, as fish, frog and newt eggs are, the water cushions and protects the embryo. These animals do not have an amnion.

Until that time in the history of life when the amnion evolved, all backboned animals had to live near water where they could lay their eggs.

The evolution of the amnion allowed animals to live all their lives on land, not even returning to water to breed.

Did dinosaurs have an amnion?

An amnion is too delicate to be preserved as a fossil. So we can never be sure which fossil animals were amniotes. But to help us decide, we can use our knowledge of living animals to interpret the remains of extinct ones.

Fossils show us that some dinosaurs laid shelled eggs on land. We know that all living backboned animals that lay their eggs on land have an amnion. So we assume that dinosaurs had an amnion too.

Protoceratops eggs.
(Smaller than lifesize.
Actual size of egg 155 mm.)

Protoceratops is a small dinosaur closely related to *Triceratops*. Its fossilized eggs and young have been found in the Gobi Desert in Mongolia.

Baby *Protoceratops* hatching.
(This is a reconstruction based on fossil embryo skeletons and eggs.)

ichthyosaur

plesiosaur

Did any other extinct animals have an amnion?

We very rarely know about the soft parts of extinct animals, because usually only their bones and teeth are preserved as fossils.

But we assume that, if extinct animals and living animals share homologues in their bones and teeth, they probably share homologues in their soft parts too.

Ichthyosaurs and **plesiosaurs** share a number of structural similarities with living animals that have an amnion, so we assume that they had an amnion too.

Ichthyosaurs were so well adapted to life in the sea that it is very unlikely that they were able to come out on to land to reproduce. Some ichthyosaurs have been preserved with their young inside them. This suggests that, like whales and dolphins, they gave birth to live young.

This ichthyosaur has been preserved with the broken-up skeletons of three unborn young inside it. A fourth young one may have been born just as the mother died – its skeleton can be clearly seen below her tail.

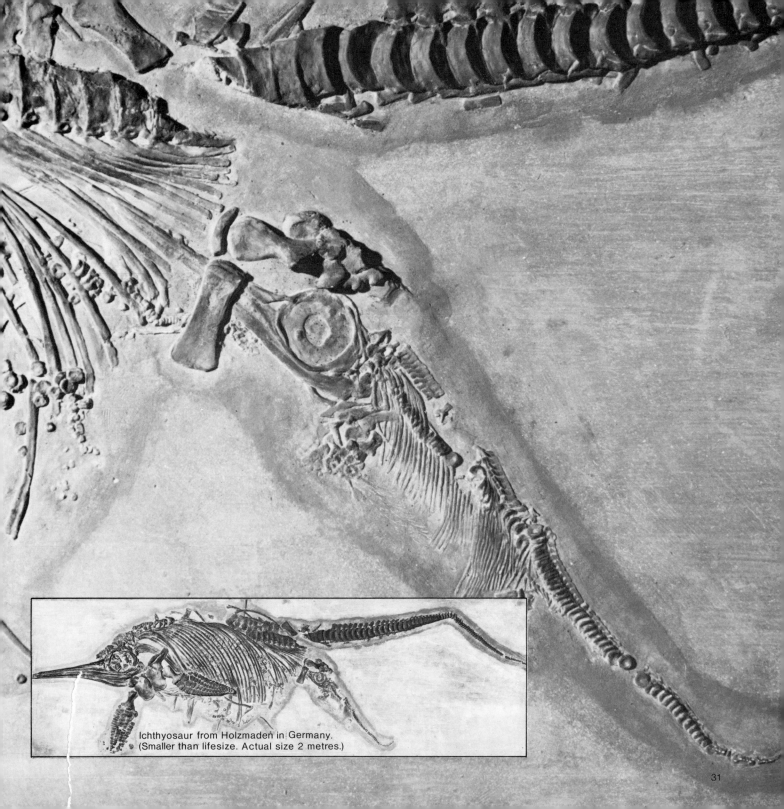

Ichthyosaur from Holzmaden in Germany.
(Smaller than lifesize. Actual size 2 metres.)

How are dinosaurs related to other amniotes?

When we compare the skulls of dinosaurs with those of other amniotes, we find that dinosaurs share a homologue with **fossil birds** and **crocodiles**.

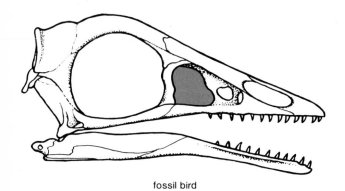

fossil bird

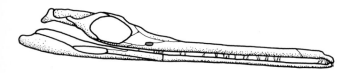

fossil crocodile

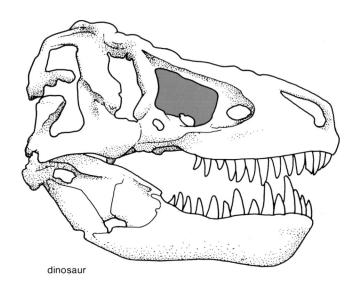

dinosaur

Extinct forms of birds and crocodiles have a hole* in the skull, in front of the eye socket.

Modern birds and crocodiles do not have this feature. But we assume that their ancestors did, and that it has since disappeared.

Dinosaurs have the same hole in the skull.

Birds, crocodiles and dinosaurs belong to a clade within the amniotes. Members of this clade are known as **archosaurs**.

* The holes are shown in red on these diagrams.

Two other types of extinct animals, **pterosaurs** and **thecodontians**, also have a hole in the skull in front of the eye socket. They, too, belong to the archosaur clade.

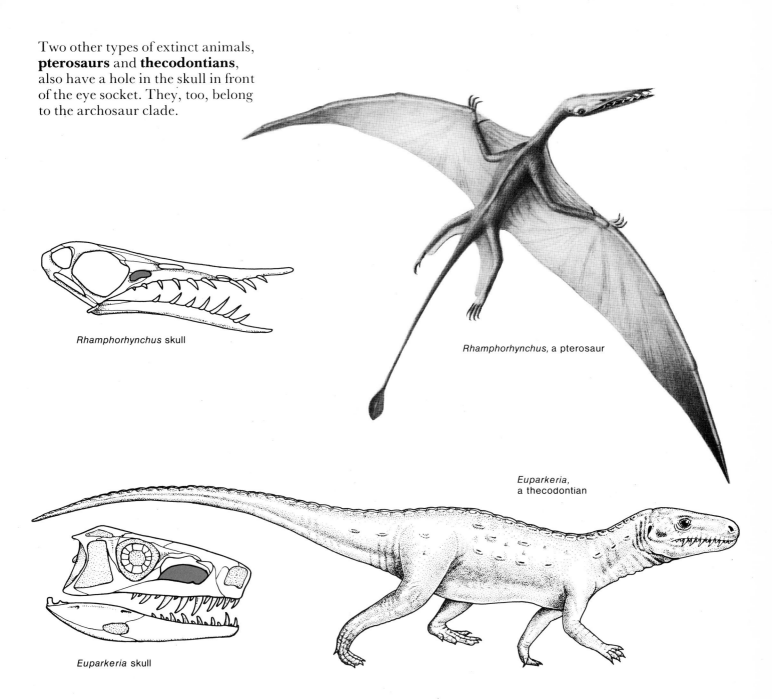

Rhamphorhynchus skull

Rhamphorhynchus, a pterosaur

Euparkeria, a thecodontian

Euparkeria skull

Chapter 4
A closer look at dinosaurs

**Dinosaurs can be distinguished
from most other archosaurs by
the way they stood.**

Varanus salvator (a monitor lizard)

Euparkeria (a thecodontian)

'Sprawlers' have to use energy
just to lift their bodies off the
ground as their legs do not carry
the body from underneath. These
animals move as if swimming on
land.

With this **'semi-improved'**
stance, the legs carry some of the
body weight from underneath and
the animal can run in short bursts.

Triceratops (a dinosaur)

Dinosaurs had what is known as a **'fully improved'** stance. Their legs supported their weight directly, and they could move easily and efficiently.

Apart from birds, no other archosaurs stand in this way.

Although all dinosaurs had a 'fully improved' stance, this feature is not a homologue.

The evolution of an upright stance is associated with changes in hip structure. When we study dinosaur hips we find that they are of *two* distinct types – there seem to have been at least two different ways of producing a 'fully improved' stance.

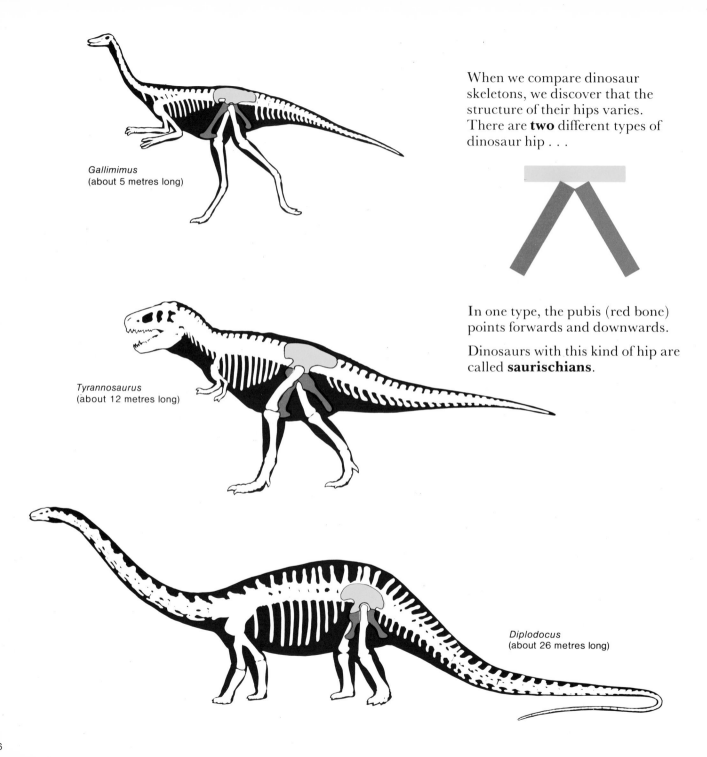

Gallimimus
(about 5 metres long)

When we compare dinosaur skeletons, we discover that the structure of their hips varies. There are **two** different types of dinosaur hip . . .

In one type, the pubis (red bone) points forwards and downwards.

Dinosaurs with this kind of hip are called **saurischians**.

Tyrannosaurus
(about 12 metres long)

Diplodocus
(about 26 metres long)

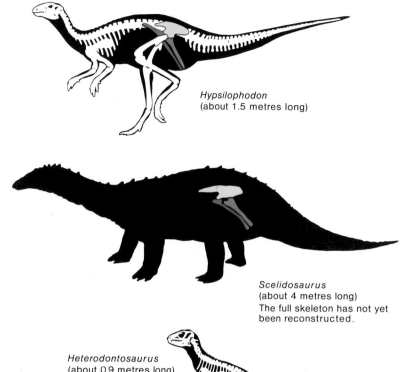

Hypsilophodon
(about 1.5 metres long)

In the other type, the main part of the pubis points backwards and downwards*, and lies parallel with the ischium (blue bone).

Dinosaurs with this kind of hip are called **ornithischians**.

Scelidosaurus
(about 4 metres long)
The full skeleton has not yet been reconstructed.

Heterodontosaurus
(about 0.9 metres long)

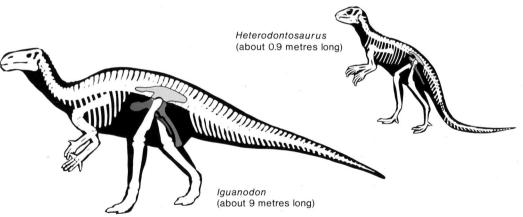

Iguanodon
(about 9 metres long)

* In some ornithischians, such as *Iguanodon*,
  the pubis also has a forward-pointing prong.

Do saurischian dinosaurs form a clade?

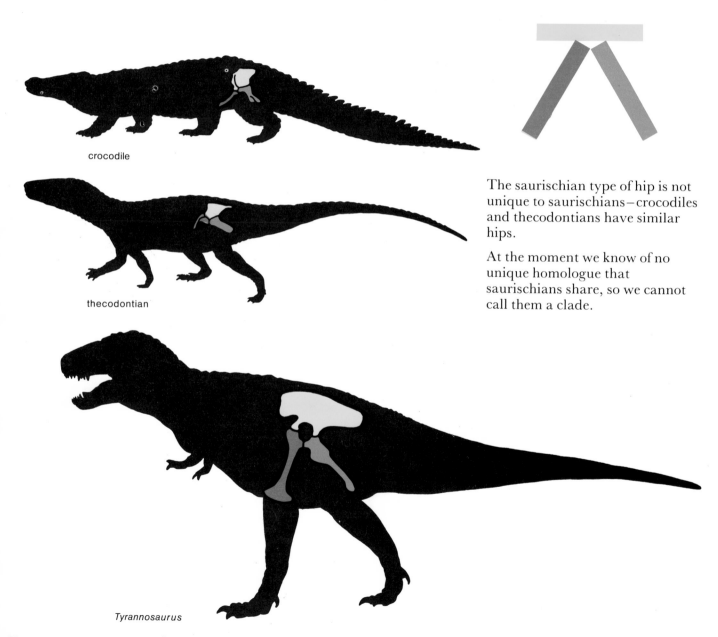

crocodile

thecodontian

Tyrannosaurus

The saurischian type of hip is not unique to saurischians—crocodiles and thecodontians have similar hips.

At the moment we know of no unique homologue that saurischians share, so we cannot call them a clade.

Do ornithischian dinosaurs form a clade?

This type of hip is unique to ornithischians*.

Ornithischians share other unique homologues. For example, they all have a **predentary bone**† in the lower jaw – no other archosaur has one.

So ornithischians do form a clade.

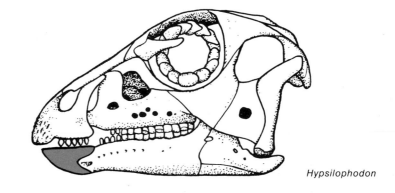

Hypsilophodon

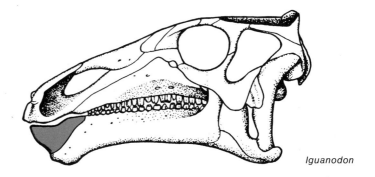

Iguanodon

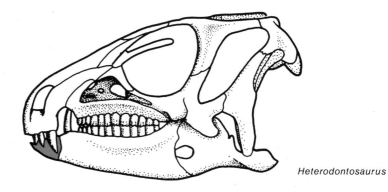

Heterodontosaurus

* Ornithischian means 'bird-hipped'. The dinosaurs were so named because their hips are similar to bird hips. However, this similarity is only superficial and is therefore misleading. (See page 48.)

† The predentary bone is shown in red on the diagrams.

How are these ornithischians related to each other?

Their relationship can be represented on a cladogram* like this. Can you put the four dinosaurs in their correct positions on the cladogram?

First: Find a feature that is shared by only two of the dinosaurs. (Clue 1, on page 41, will help you.)

Next: Now you must find another feature, which links just *one* of the remaining two dinosaurs to these two. (Refer to clue 2 if you get stuck.)

Check your answers on page 42.

* Go back to page 20, if you want to be reminded about what cladograms represent.

Iguanodon

Scelidosaurus

Hypsilophodon

Heterodontosaurus

Clue 1: Look for the two dinosaurs whose red bones have a forward-pointing prong.

Clue 2: Look at the way the dinosaurs stand.

This cladogram shows how the four ornithischians are related . . .

Check your solution against the cladogram.

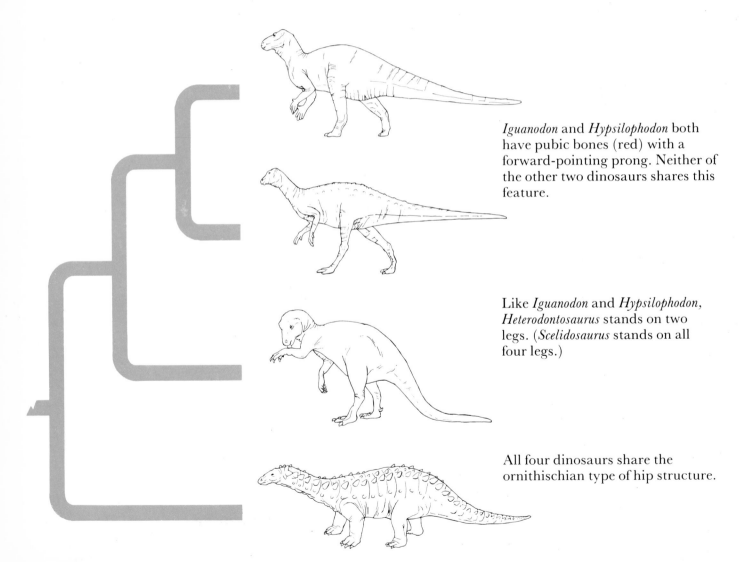

Iguanodon and *Hypsilophodon* both have pubic bones (red) with a forward-pointing prong. Neither of the other two dinosaurs shares this feature.

Like *Iguanodon* and *Hypsilophodon*, *Heterodontosaurus* stands on two legs. (*Scelidosaurus* stands on all four legs.)

All four dinosaurs share the ornithischian type of hip structure.

How are these three saurischians related to each other?

Two of the dinosaurs share features not shared by the third. *Can you spot them?*

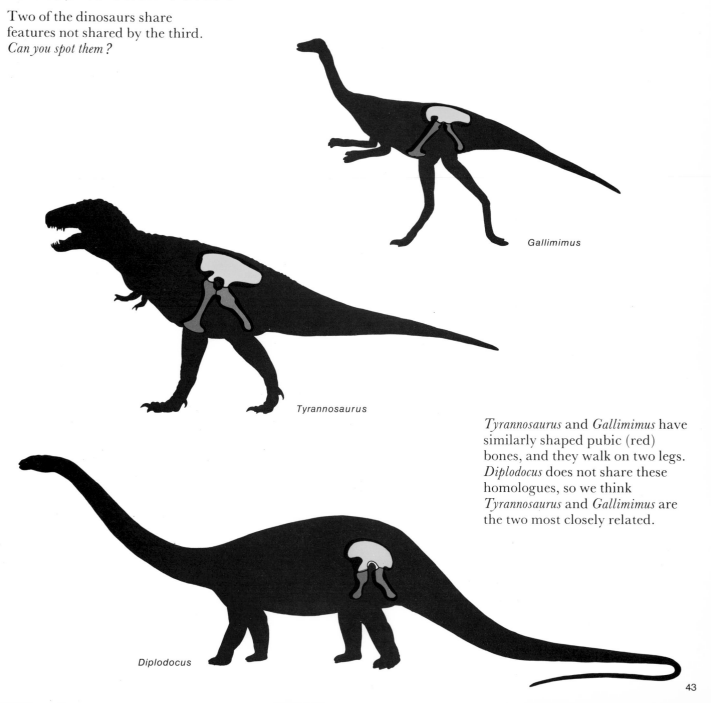

Gallimimus

Tyrannosaurus

Diplodocus

Tyrannosaurus and *Gallimimus* have similarly shaped pubic (red) bones, and they walk on two legs. *Diplodocus* does not share these homologues, so we think *Tyrannosaurus* and *Gallimimus* are the two most closely related.

Chapter 5
Birds – the dinosaurs' closest living relatives?

Birds are archosaurs – and are thus related to dinosaurs. In this chapter we will look at evidence that suggests that birds may be the **closest living relatives** of the dinosaurs.

As you saw on page 20, birds form a clade. All birds have **feathers**.

The **wishbone*** is another feature unique to birds. However, not all birds have a wishbone. It is absent in many flightless birds such as ostriches and the extinct moas. And in emus it is very small.

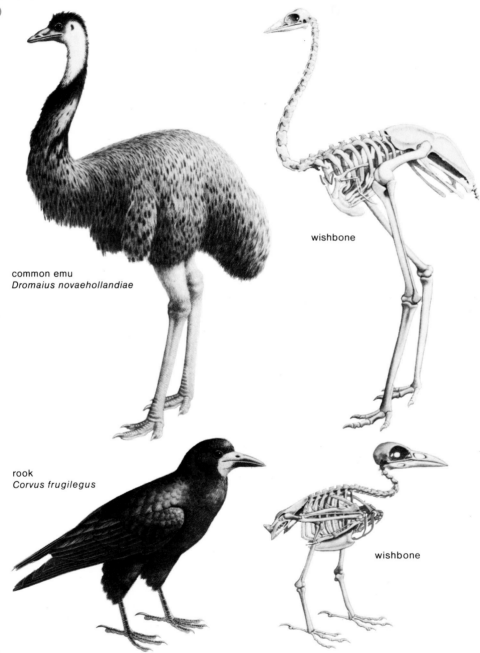

common emu
Dromaius novaehollandiae

wishbone

rook
Corvus frugilegus

wishbone

* This bone is thought to have been formed by the two collar bones fusing together.

This is *Archaeopteryx* – a fossil bird that lived almost 150 million years ago. *Archaeopteryx* is classified as a bird, because it has **feathers** and a **wishbone**. You can see these features on this fossil.

◆ feathers
● wishbone

The London *Archaeopteryx*

This fossil was the first *Archaeopteryx* to be discovered. It was found in 1861 inside a limestone slab in a German quarry, and is now in the British Museum (Natural History).

Finding the birds' closest relatives

You have seen that, to work out relationships between groups of animals, we must look for features that the animals have in common.

The fossil bird *Archaeopteryx* shares features with other archosaurs that modern birds no longer possess*:

- a long bony tail
- claws on the fingers
- teeth.

These features are labelled on the Berlin fossil.

In the rest of this chapter, we will look for other features that *Archaeopteryx* shares with archosaurs. We will compare *Archaeopteryx* with each of the other major archosaur groups in turn:

- pterosaurs
- ornithischian dinosaurs
- saurischian dinosaurs
- the early archosaurs
- crocodiles.

If we find a feature that *Archaeopteryx* shares with only one of these groups, we may have found the birds' closest relatives.

The Berlin *Archaeopteryx*

This fossil was discovered in Germany in 1877. It is now in the Museum of Natural History, East Berlin.

▲ tail
■ teeth
◗ claws

* We believe that, during the course of their evolution, birds lost these characteristics as they became better adapted for flight.

Is *Archaeopteryx* most closely related to pterosaurs?

They both have wings . . . but the arrangement of the bones that support the wing is very different. Also, *Archaeopteryx*'s wing has feathers, but the pterosaur's wing is covered with skin.*

The wings are not similar enough to suggest that birds and pterosaurs are very closely related.

wing of *Archaeopteryx*

wing of *Rhamphorhynchus*, a pterosaur

* A pterosaur's wing was used for gliding, whereas we now believe that *Archaeopteryx* was capable of flapping flight.

Is *Archaeopteryx* most closely related to ornithischian dinosaurs?

Archaeopteryx has hip bones that are similar in shape to those of ornithischians. (The pubic (red) bone points backwards.) Indeed, 'ornithischian' means bird-hipped, and scientists used to think that they were very closely related. But today, most scientists agree that the similarity is only superficial, and that ornithischians are *not* closely related to birds.

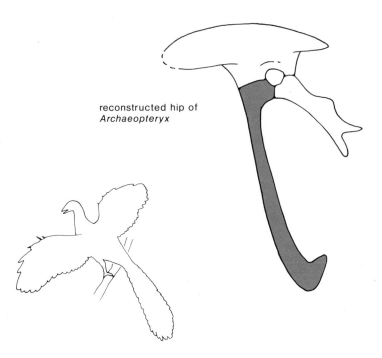

reconstructed hip of
Archaeopteryx

hip of *Iguanodon*

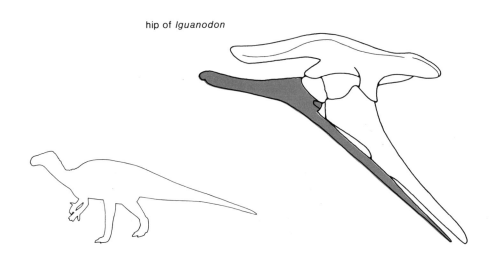

Is *Archaeopteryx* most closely related to saurischian dinosaurs?

This is *Deinonychus*, a member of a group of saurischian dinosaurs called theropods. Theropods walked on two legs and most ate meat.

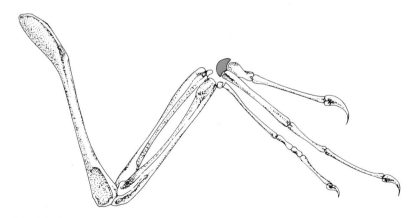

wrist of *Deinonychus*

Deinonychus and *Archaeopteryx* have similar moon-shaped bones in their wrists (shown in red on the diagrams). Many scientists believe that this feature is a unique homologue, not shared by other archosaurs. So they claim that dinosaurs such as *Deinonychus* are the birds' closest relatives.

wrist of *Archaeopteryx*

49

Not everyone agrees that saurischians and birds are so closely related – and there are two other archosaur groups to consider . . .

Early archosaurs

At present we know of no unique homologue that birds share with early archosaurs. But research on some exciting new dinosaur finds may throw light on the relationship between these groups.

Euparkeria, an early archosaur

Crocodiles

Some people believe that there are unique similarities between bones in the skulls of crocodiles and birds. This research, too, is continuing so we still do not know how birds and crocodiles are related.

Nile crocodile
Crocodylus niloticus

Debate about the relationship between birds and other archosaurs is based both on new fossil finds and on the re-examination of the fossils we already have.

As more research is carried out, and as more fossils are discovered, our ideas of the relationships between the different archosaur groups may change. We may have to reassess features which we previously thought were homologues, in the light of other evidence. Just as we no longer think that the hip bones of birds and ornithischians are really similar, so we may change our ideas about the similarities between saurischians and birds.

We cannot go back in time to find out about dinosaurs and their relatives, so we shall probably always be arguing about them!

Fossilized footprint of a dinosaur, from a Purbeck Stone quarry near Swanage in Dorset. About 135 million years old.

'Claws'

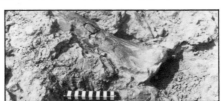

In January 1983, Mr Bill Walker, an amateur fossil hunter, came across a huge claw bone in a Surrey claypit. The claw measured 310 mm along the outside edge—half as big again as the largest of *Tyrannosaurus*'s claws.

This exciting discovery led a team of palaeontologists from the British Museum (Natural History) to excavate the site—and eventually to reveal the remains of a dinosaur. In the process, three van-loads of rock-encased bone were transported to the Museum for study.

From the beginning of the work, it was clear that a new species of dinosaur had been discovered. 'Claws', as the dinosaur was quickly nicknamed, was a bipedal, flesh-eating dinosaur that lived about 125 million years ago. It probably stood about 4 metres high (smaller than was at first expected from the size of the claw).

The find of the century

Carnivorous dinosaurs are very rare. This is the first major British find for over a century. And large carnivorous dinosaurs of this particular period are *especially* rare, anywhere in the world. A find like this reminds us how much is still waiting to be discovered about dinosaurs.

The fossil animals featured in this book

Scelidosaurus

Scelidosaurus was an ornithischian dinosaur. Only two specimens have so far been discovered—both in England. They are now in the British Museum (Natural History).

Scelidosaurus was a heavily built dinosaur, with an armour of solid bony plates along the length of its body. It was herbivorous.

Length   about 4 metres

Age   about 185 million years

Period   Lower Jurassic

	MILLIONS OF YEARS AGO
CAENOZOIC	0
	65
UPPER CRETACEOUS	
LOWER CRETACEOUS	136
UPPER JURASSIC	
LOWER JURASSIC	195
UPPER TRIASSIC	
LOWER TRIASSIC	225
UPPER PERMIAN	
LOWER PERMIAN	280
AGE OF THE EARTH	4600

Heterodontosaurus

Heterodontosaurus was an early ornithischian dinosaur. It was small and lightly built, and ate plant material.

Heterodontosaurus had unusually varied teeth. In fact, its name means 'different-toothed lizard'. The top front teeth are sharp and pointed, and bite against a horny beak on the lower jaw. The tall, ridged back teeth are packed very close together, to cut the plant material. Most unusually, *Heterodontosaurus* also has a pair of small tusks on each jaw.

Length about 0.9 metres

Age 188 million years

Period Lower Jurassic

(You can see a drawing of this skull on page 39.)

CAENOZOIC		0
UPPER CRETACEOUS		65
LOWER CRETACEOUS		
		136
UPPER JURASSIC		
LOWER JURASSIC		
		195
UPPER TRIASSIC		
LOWER TRIASSIC		225
UPPER PERMIAN		
LOWER PERMIAN		
		280
AGE OF THE EARTH		
		4600

MILLIONS OF YEARS AGO

Scolosaurus

Scolosaurus was an ornithischian dinosaur. In life, it was covered in an armour of thick bony plates, set close together in its leathery skin. The plates on its back had spikes and there were two very large spikes on the tip of its tail. Its armour would have protected it against flesh-eating dinosaurs.

Scolosaurus probably fed on soft plant material—it had weak teeth.

Length up to 6 metres

Age 73 million years

Period Upper Cretaceous

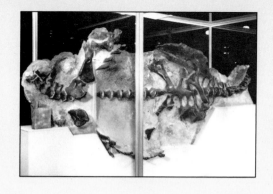

	MILLIONS OF YEARS AGO
CAENOZOIC	0
	65
UPPER CRETACEOUS	
	136
LOWER CRETACEOUS	
UPPER JURASSIC	
LOWER JURASSIC	195
UPPER TRIASSIC	
LOWER TRIASSIC	225
UPPER PERMIAN	
LOWER PERMIAN	
	280
AGE OF THE EARTH	4600

Triceratops

Triceratops was a rhinoceros-like ornithischian dinosaur. It probably roamed in large herds, grazing the vegetation. It would have used its horny beak to cut up plants, and then ground them with its rows of strong back teeth. Its horns and the large bony frill over its neck would have protected it against flesh-eating dinosaurs such as *Tyrannosaurus*.

Length about 7 metres

Age 70–65 million years

Period Upper Cretaceous

	MILLIONS OF YEARS AGO
CAENOZOIC	0
	65
UPPER CRETACEOUS	
LOWER CRETACEOUS	136
UPPER JURASSIC	
LOWER JURASSIC	195
UPPER TRIASSIC	
LOWER TRIASSIC	225
UPPER PERMIAN	
LOWER PERMIAN	280
AGE OF THE EARTH	4600

Iguanodon

Iguanodon was an ornithischian dinosaur. Its fossilized footprints indicate that it normally walked on two legs. But it has hoof-like bones on its hands which suggest that it sometimes walked on all fours. The 'thumb-spikes' on its hands could have been used for defence.

Iguanodon fed on plants. At the front of its mouth it had a horny beak instead of teeth. It would have used its beak to cut up the plants and then ground them with the teeth at the back of its mouth.

Length   about 9 metres

Age   140–105 million years

Period   Lower Cretaceous

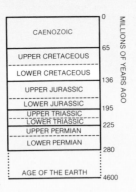

	MILLIONS OF YEARS AGO
CAENOZOIC	0
	65
UPPER CRETACEOUS	
LOWER CRETACEOUS	
	136
UPPER JURASSIC	
LOWER JURASSIC	
	195
UPPER TRIASSIC	
LOWER TRIASSIC	225
UPPER PERMIAN	
LOWER PERMIAN	
	280
AGE OF THE EARTH	
	4600

Hypsilophodon

Hypsilophodon was a fast-moving, plant-eating ornithischian dinosaur. We think it was able to run fast because the lower parts of its legs were much longer than the upper parts—just as in fast-moving animals alive today. *Hypsilophodon* had a stiffened tail that was probably held out straight behind to help it balance as it ran. Running away was probably its best defence against enemies.

Length   about 1.5 metres

Age   about 115 million years

Period   Lower Cretaceous

		MILLIONS OF YEARS AGO
CAENOZOIC		0
UPPER CRETACEOUS		65
LOWER CRETACEOUS		
UPPER JURASSIC		136
LOWER JURASSIC		195
UPPER TRIASSIC		
LOWER TRIASSIC		225
UPPER PERMIAN		
LOWER PERMIAN		280
AGE OF THE EARTH		4600

	MILLIONS OF YEARS AGO
CAENOZOIC	0
	65
UPPER CRETACEOUS	
LOWER CRETACEOUS	136
UPPER JURASSIC	
LOWER JURASSIC	195
UPPER TRIASSIC	
LOWER TRIASSIC	225
UPPER PERMIAN	
LOWER PERMIAN	280
AGE OF THE EARTH	4600

Diplodocus

Diplodocus was a saurischian dinosaur. It is one of the largest land animals that has ever lived. It was about 26 metres long and would have weighed about 10 tonnes.

Scientists used to think that *Diplodocus* lived in lakes or swamps, but we now think that this is unlikely for the following reasons:

- Most large animals that live in water (e.g. hippopotamuses) have a barrel-shaped body with a short neck and legs. *Diplodocus* is not this shape.

- *Diplodocus* has small feet in comparison with the size of its body, so it would have found it difficult to walk on soft, swampy ground without sinking in.

- The rocks where *Diplodocus* was found contain the remains of land plants and animals, and not swamp-living ones.

We now think *Diplodocus* probably lived on land.

Length	about 26 metres
Age	about 150 million years
Period	Upper Jurassic

Gallimimus

Gallimimus was a slender, ostrich-like saurischian dinosaur with long powerful back legs for running fast. Some scientists think *Gallimimus* fed on small animals such as insects and lizards which it snapped up in its horny beak. Others suggest that it fed on plants or possibly even on the eggs of other dinosaurs.

Its three-fingered hands could have been used to tear up prey, but it seems just as likely that they were used for digging or searching amongst the vegetation for food.

Length   about 5 metres

Age   70–65 million years

Period   Upper Cretaceous

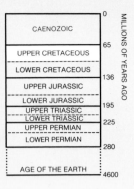

	MILLIONS OF YEARS AGO
CAENOZOIC	0
	65
UPPER CRETACEOUS	
LOWER CRETACEOUS	136
UPPER JURASSIC	
LOWER JURASSIC	195
UPPER TRIASSIC	
LOWER TRIASSIC	225
UPPER PERMIAN	
LOWER PERMIAN	280
AGE OF THE EARTH	4600

MILLIONS OF YEARS AGO	
CAENOZOIC	0
	65
UPPER CRETACEOUS	
LOWER CRETACEOUS	
	136
UPPER JURASSIC	
LOWER JURASSIC	
	195
UPPER TRIASSIC	
LOWER TRIASSIC	225
UPPER PERMIAN	
LOWER PERMIAN	
	280
AGE OF THE EARTH	
	4600

Tyrannosaurus

Tyrannosaurus was a saurischian dinosaur. It is the largest known flesh-eating land animal that has ever lived. It was 12 metres long, 5 metres high and would have weighed up to 7 tonnes. Its long teeth had serrated edges for tearing flesh.

Tyrannosaurus had large, powerful back legs, with clawed feet that were probably used to grip its prey. It is difficult to imagine what its tiny front limbs were used for— they were so small they would not even have reached its mouth.

Length	about 12 metres
Age	70–65 million years
Period	Upper Cretaceous

Deinonychus

Deinonychus was a small saurischian dinosaur. It was carnivorous and used its long, clawed hands to grasp its prey.

The most remarkable feature of this dinosaur was the tremendous claw on the second toe of each hind foot. The name '*Deinonychus*' means 'terrible claw'.

The tail was strengthened by long bony rods. It was probably held out horizontally, to balance the dinosaur as it ran or struck out with a clawed foot.

Length   about 2.5 metres

Age   about 115 million years

Period   Lower Cretaceous

MILLIONS OF YEARS AGO	
CAENOZOIC	0
UPPER CRETACEOUS	65
LOWER CRETACEOUS	136
UPPER JURASSIC	
LOWER JURASSIC	195
UPPER TRIASSIC	
LOWER TRIASSIC	225
UPPER PERMIAN	
LOWER PERMIAN	280
AGE OF THE EARTH	4600

Ichthyosaurus

Ichthyosaurus is not a dinosaur. It is an ichthyosaur. You can see from its shape that it was suited to life in water—it was streamlined, had fin-like paddles instead of legs, a tail fin and a fin on its back.

Ichthyosaurus lived all its life in the sea and probably fed on fish or shellfish (ammonites), which it caught with its sharp teeth. It had no gills so it could not take in oxygen from the water and, like dolphins and porpoises, must have come to the surface to breathe.

Ichthyosaurus did not even come on land to lay its eggs—it gave birth to live young in the water.

Length   up to 12 metres

Age   about 185 million years

Period   Lower Jurassic

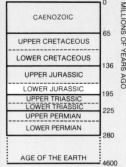

	MILLIONS OF YEARS AGO
CAENOZOIC	0
UPPER CRETACEOUS	65
LOWER CRETACEOUS	136
UPPER JURASSIC	
LOWER JURASSIC	195
UPPER TRIASSIC	
LOWER TRIASSIC	225
UPPER PERMIAN	
LOWER PERMIAN	280
AGE OF THE EARTH	4600

Plesiosaurus

Plesiosaurus is not a dinosaur. It is a plesiosaur. It lived in water and would have used its paddle-like legs to swim quickly. It probably fed on fish which it snapped up with its sharp pointed teeth.

We do not know whether plesiosaurs came ashore to lay eggs or whether, like ichthyosaurs, they gave birth to live young in the water. They may have been able to move about on land, but only in a clumsy way like seals.

Length   up to 15 metres

Age   about 180 million years

Period   Lower Jurassic

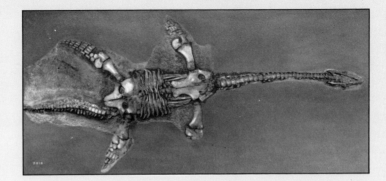

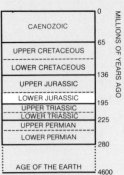

	MILLIONS OF YEARS AGO
CAENOZOIC	0
UPPER CRETACEOUS	65
LOWER CRETACEOUS	136
UPPER JURASSIC	
LOWER JURASSIC	195
UPPER TRIASSIC	
LOWER TRIASSIC	225
UPPER PERMIAN	
LOWER PERMIAN	280
AGE OF THE EARTH	4600

Pteranodon

Pteranodon is not a dinosaur. It is a pterosaur. It is one of the largest known flying animals that has ever lived—it had a wingspan of 7 metres. Because it had hollow bones, *Pteranodon* was very light for its size—adults weighed less than 17 kilograms (about the same weight as a large turkey).

Pteranodon probably flew by gliding with its large wings stretched out. It would have been very clumsy on the ground because it had weak legs and could not fold its wings completely, as bats and birds can.

Some scientists have suggested that *Pteranodon* lived on cliffs where it could take off into rising air currents. Its bones have been found in rocks formed from sea sediments, so it probably glided across the sea and swooped down to catch fish in its long toothless beak.

Wingspan	up to 7 metres
Age	about 80 million years
Period	Upper Cretaceous

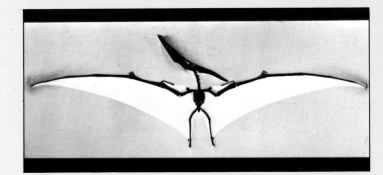

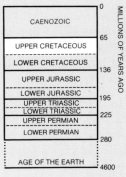

	MILLIONS OF YEARS AGO
CAENOZOIC	0
	65
UPPER CRETACEOUS	
LOWER CRETACEOUS	136
UPPER JURASSIC	
LOWER JURASSIC	195
UPPER TRIASSIC	
LOWER TRIASSIC	225
UPPER PERMIAN	
LOWER PERMIAN	280
AGE OF THE EARTH	4600

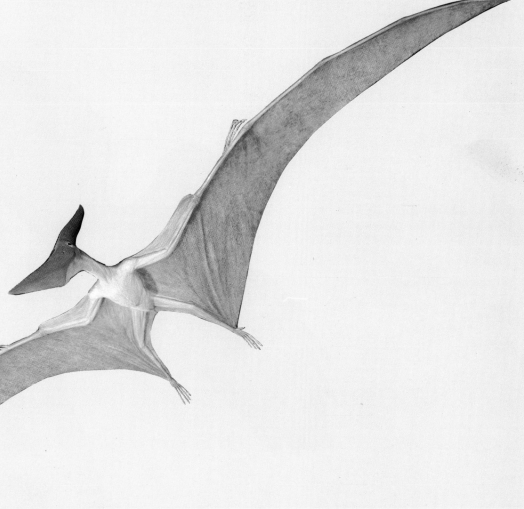

Archaeopteryx

Archaeopteryx is the earliest known bird. We can see from its fossil remains that it had a wishbone and feathers, two features unique to birds.

There are two ideas about the way *Archaeopteryx* lived. The traditional idea is that it lived in trees, and glided or flew from branch to branch. The other idea is that *Archaeopteryx* lived on the ground, and used to run along with its feathery 'arms' outstretched – perhaps to trap insects. As it ran along like this, it might have been able to lift off from the ground and fly.

It now seems certain that *Archaeopteryx* was capable of flapping flight, but we are not sure how well it could fly. When modern birds flap their wings, they use very large wing muscles, some of which attach to a keel-shaped breastbone. *Archaeopteryx* does not have a breastbone like this, and its wing muscles may have been quite small.

Wingspan about 0.5 metres

Age about 147 million years

Period Upper Jurassic

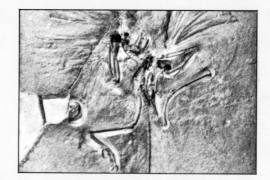

	MILLIONS OF YEARS AGO
CAENOZOIC	0
	65
UPPER CRETACEOUS	
LOWER CRETACEOUS	136
UPPER JURASSIC	
LOWER JURASSIC	195
UPPER TRIASSIC	
LOWER TRIASSIC	225
UPPER PERMIAN	
LOWER PERMIAN	280
AGE OF THE EARTH	4600

Reconstructing a dinosaur

Making a model dinosaur involves . . .

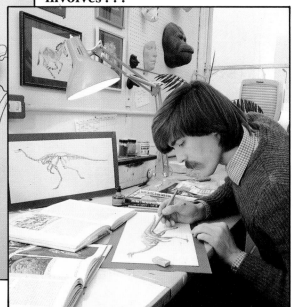

using what you know from the dinosaur remains . . .

First of all, the model-maker measures each of the bones carefully, and makes an accurate scale drawing of the dinosaur's skeleton. Clues such as the shape of the limbs and feet can tell a great deal about how the animal stood. These scale drawings are used to construct the framework of the model.

Stages in the reconstruction of *Gallimimus* at the British Museum (Natural History).

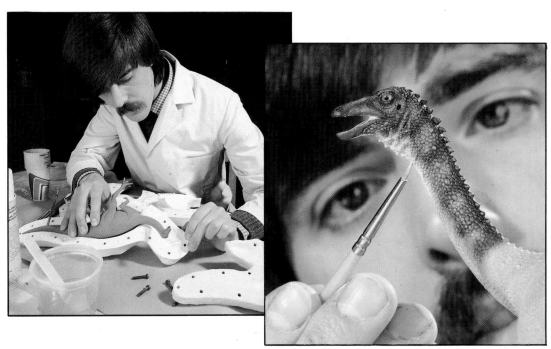

using your knowledge of present-day animals . . .

The dinosaurs' closest living relatives—birds and crocodiles—can help with the next stage . . . By comparing the skeletons, the model-maker can work out how the muscles on the dinosaur's skeleton might have been arranged. Clay can then be added to give shape to the model.

and using your imagination . . .

Details such as skin texture and colour can only be guessed at. We have *some* fossil impressions of dinosaur skin on which to base the decision, but the colour is anyone's guess!

Glossary

Amnion: a membrane surrounding the embryo of certain animals. It contains a fluid that keeps the embryo wet and cushions it against bumps and knocks.

Amniotes: a clade that includes birds, crocodiles, lizards and snakes, turtles and tortoises, mammals, dinosaurs, ichthyosaurs, plesiosaurs, pterosaurs, and thecodontians. The unique homologue that these animals share is an amnion.

Analogue: a feature shared by two or more living things which is only superficially similar, and does not indicate any common ancestry.

Archosaurs: a clade that includes birds, crocodiles, dinosaurs, pterosaurs and thecodontians. The unique homologue that these animals share is a particular hole in the skull in front of the eye.

Clade: a group comprising all those living things that share a unique homologue. In evolutionary terms, a clade comprises all the descendants of a single common ancestor.

Cladogram: a branching diagram that represents the relationship between different clades. (See page 22.)

Common ancestor: an ancestor shared by two or more living things.

Fossil: the remains of a living thing, or direct evidence of its presence, preserved in rocks. Usually only hard parts such as bones, teeth and shells are preserved.

Homologue: a feature that defines a clade. In evolutionary terms, a homologue is a feature that is similar in two or more organisms because they inherited it from a common ancestor.

Ornithischian dinosaurs: one of the two types of dinosaur. They are distinguished by having a hip shaped like this:

The group includes dinosaurs such as *Heterodontosaurus*, *Scelidosaurus*, *Iguanodon* and *Hypsilophodon*. It is a clade because ornithischians share a unique homologue – they have a predentary bone in the lower jaw. (See page 39.)

Saurischian dinosaurs: one of the two types of dinosaur. They are distinguished by having a hip shaped like this:

Saurischian means 'lizard-hipped'. The group includes dinosaurs such as *Diplodocus*, *Tyrannosaurus*, *Gallimimus* and *Deinonychus*. We know of no unique homologue that saurischians share, so we cannot call the group a clade.

Acknowledgements for photographs
16: wild sheep, Hans Reinhard/Bruce Coleman Ltd; Madagascar flying fox, Heather Angel.
18: bluebottle, Stephen Dalton/Bruce Coleman Ltd; southern right whale, J Bartlett/ Bruce Coleman Ltd; Madagascar flying fox, Heather Angel. **28:** frogs and spawn, Frank Greenaway.

Further reading

. . . about dinosaurs

Dinosaurs by Anne McCord, *The Children's Prehistory* series, Usborne 1977. For younger children.

A New Look at the Dinosaurs by Alan Charig, Heinemann/British Museum (Natural History) 1983 (reprinted with amendments). For older children and adults. An up-to-date account illustrated with many new and scientifically accurate drawings. Specially recommended.

Dinosaurs by L. B. Halstead and J. Halstead, Blandford Press 1981.

A Natural History of Dinosaurs by R. T. J. Moody, Hamlyn 1977.

The World of Dinosaurs by M. Tweedie, Weidenfeld & Nicolson 1977. Also deals with ichthyosaurs, plesiosaurs, crocodiles and pterosaurs.

Men and Dinosaurs: The search in field and laboratory by E. H. Colbert, Evans 1968. The history of dinosaur collecting in North America and Europe.

Collins Guide to Dinosaurs by D. Lambert, Collins 1983. A 'field guide' to dinosaurs, giving general and specific information on over 300 species, together with maps.

The Age of Reptiles by E. H. Colbert, Weidenfeld & Nicolson 1965.

Life before Man by Z. V. Spinar and Z. Burian, Thames & Hudson 1972. Deals with fossil amphibians, reptiles, birds and mammals.

. . . about fossil birds

The Age of Birds by Alan Feduccia, Harvard University Press 1980.

. . . about evolution

Evolution by Colin Patterson, British Museum (Natural History) 1978. An introduction to all aspects of evolutionary theory, written especially for those with little or no knowledge of biology.

. . . about classification

Classification – a beginner's guide to some of the systems of biological classification in use today, British Museum (Natural History) 1983. Compares cladistics with other systems of classification.

. . . about living reptiles

The Life of Reptiles by A. d'A. Bellairs, 2 volumes, The Weidenfeld & Nicolson Natural History Library 1969. Also deals with fossil reptiles.

Reptiles by A. d'A. Bellairs and J. Attridge, Hutchinson University Library 1975. Also deals with fossil reptiles.

The World of Amphibians and Reptiles, Sampson Low Guides 1978.

Living Reptiles of the World by K. P. Schmidt and R. F. Inger, Hamish Hamilton 1957.

The Reptiles by A. Carr, Time Life International 1964.

Index

Published by the British Museum (Natural History), London and the Press Syndics of the University of Cambridge
The Edinburgh Building, Shaftesbury Road, Cambridge CB2 2RU
32 East 57th Street, New York, NY 10022, USA
10 Stamford Road, Oakleigh, Melbourne, 3166, Australia

© Trustees of the British Museum (Natural History) 1979, 1985

First Published 1979
Second Edition 1985

Library of Congress Cataloging-in-Publication Data
British Museum (Natural History)
    Dinosaurs and their living relatives.
    Bibliography: p.
    Includes index.
    Summary: Describes the characteristics of dinosaurs and some of the animals, specifically birds and crocodiles, physically related to dinosaurs that are still living today.
    1. Dinosaurs.   2. Birds.   3. Crocodiles.
[1. Dinosaurs.   2. Birds.   3. Crocodiles]   I. Title.
QE862.D5B68 1985   567.9'1   85–21273
ISBN 0–521–26426–X
ISBN 0–521–26970–9 (pbk.)

Printed in Great Britain by Acolortone Ltd, Ipswic